CAREERS IN
PLUMBING

PLUMBING CONTRACTING

HOW DO FEEL ABOUT WORKING WITH your hands? About working in a different place every few days or weeks? Getting paid by the hour or by the job? Becoming self-employed as a freelancer or independent business owner? If any of these prospects appeal to you, you may have what it takes to create a successful career in plumbing.

Why plumbing? It is a perfect career for people who like to work with their hands, are not keen on being stuck behind a desk, and aspire to be their own boss someday. Plumbing is necessary to life in the modern world. Can you even imagine a world without plumbing? No running water or flush toilets? You have come to take such things for granted. Most of the world's people manage to live without them, but suffer severe public health consequences as a result. Plumbing may seem mundane but it is actually one of the keys to building and maintaining a healthy, modern society.

You have picked an excellent time to start researching careers in plumbing. Environmental awareness has increased the demand for new and more-efficient plumbing systems, thereby boosting the need for skilled plumbers to build and maintain them. Plumbing is also relatively recession-proof, as even old plumbing systems must be maintained regardless of larger economic conditions. When the economy is depressed people may cut back building new houses, but they still have to keep their kitchens and bathrooms functioning.

Pay close attention to the information contained in this report. In it you will find sections on what kind of education you will need to start your career, how to get some experience before committing to advanced training, and even what you may like and dislike about the career. If you are intrigued by what you read here, keep going. Be sure to check out the links on the last page of this report. You can never know too much.

WHAT YOU CAN DO NOW

YOU CAN GET A LOOK INSIDE PLUMBING careers right now. Take a few shop classes while you are still in high school. Learn how to use tools! It does not really matter what tools you use, just get started. Plumbing is what is known as a skilled trade, a broad career classification that also includes such professions as carpenters, electricians, and heating and air conditioning professionals. All of these careerists use a wide variety of hand and power tools, and there is plenty of overlap in basic skills. A class in machine shop will teach you about metalworking, which is important for plumbers to know. A class in automotive technology will teach you to use an enormous variety of tools, and a class in carpentry will teach you all about the need to be precise. All of these classes will also teach you about safety on the job, which is important for everybody.

Have you ever taken a hard look at the pipes in your own home? Spend some time crawling around under the sinks in your home. Follow the water lines to the basement or wherever they enter the home. Get a do-it-yourself manual and learn how to tackle simple plumbing jobs like unclogging drains or changing faucet cartridges. You might be surprised at how many plumbing jobs you can do yourself.

There are jobs you cannot do yourself. For that you will need to hook up with an actual plumber. All states require licensed plumbers to be at least 18 years old – a technicality because apprenticeship programs also require new apprentices to be at least 18 years old and most take four to five years to complete – but there is no law prohibiting an aspiring careerist of any age from helping a plumber do simple things like haul tools from one work site to the next. Many plumbers are self-employed and are very approachable. Find a few in your neighborhood and ask if they could use an extra set of hands for a few hours a week.

HISTORY OF THE CAREER

MOST PEOPLE CONSIDER PLUMBING to be a modern convenience, a pleasant, hygienic alternative to the outhouses of not-too-many years ago. The quest for efficient, sanitary plumbing has been at the top of humankind's wish list for thousands of years.

The earliest plumbing, dating back tens of thousands of years to human's first attempts at agriculture, took the form of irrigation, or diverting water from where it was naturally flowing to where it needed to be to water crops. Archaeological evidence from all over the world shows that ancient people dug simple trenches to divert water from streams and rivers into their plots to help plants grow. This was not quite plumbing as we know it today, but it was the first attempt to take control over a source of water, which is ultimately what plumbing is all about.

By about 6500 BC, humans started to dig wells to find water. Digging a well meant people did not necessarily have to live near an open water source like a river, lake or stream. Villages spring up around wells, and wells served a secondary social function as residents chatted with each other during their daily trips to fill water jugs to carry back to their homes.

Wells were not sufficient for the first cities, however. The cities of the Roman Empire, for example, made use of aqueducts to move water over long distances and into the cities where it was needed. Many Roman aqueducts can still be seen today all across Europe. They no longer carry water but the fact that they are still standing is ample testimony to how much importance the Romans placed on their construction. Many Roman cities show evidence of complex plumbing systems. In Pompeii, for example, visitors can see public and private toilet facilities, fountains, bathing facilities and gutters and culverts to whisk away sewage. Romans used lead to fashion pipes to bring water into individual homes and businesses, many of which can still be seen. Although it is easy to roll into pipes, lead is also quite toxic after long exposure and can lead to brain damage. It has even been suggested that lead pipes played a role in the downfall of the Roman Empire by slowly poisoning upper-class Romans who had lead-pipe plumbing in their homes.

Most people may think of plumbing as being primarily about washing, bathing, cooking and cleaning. Its most important function, however, is sanitation. A toilet is a comfortable convenience, but have you ever considered a world without toilets? Open defecation is the single-largest hazard to public health worldwide. Human feces attracts rats and rats carry all manner of diseases. The great plagues of medieval Europe were caused largely by the lack of proper sanitation. Epidemics still break out in parts of India today, where only about half the population uses modern flush toilets connected to municipal sewer systems.

As cities grew, plumbing had to become more complex and less expensive so that it could reach more people. Many private water companies were established in London in the early 1700s to bring clean water and sanitation to crowded neighborhoods. Plumbing technology continued to improve over time, significantly improving the quality of life for the residents of the world's first metropolis. One of the great innovators in indoor plumbing technology in the late 1800s was an inventor named Thomas Crapper, who held nine patents related to bathroom technology. Whether or not he also lent his name to the process is still debated by historians.

The history of plumbing since the early 1900s has mostly been about enhancing efficiency and access. A century ago many Americans still used outhouses. Virtually none do today. Public sanitation is so efficient that nobody gives it much thought. Drinking water is filtered and lightly chlorinated to kill germs, sewage is washed away underground and treated in enormous plants that give off very little waste, and many Americans now get their drinking water from a chilled spout in their refrigerator door. The progress in plumbing over the centuries has been amazing!

For plumbers, plumbing has never been more interesting. For at least a century cast-iron pipe was the standard for plumbing in most jurisdictions. It worked well but was difficult to move around and even more difficult to cut. Today, most jurisdictions allow plumbers to use PVC pipe – polyvinyl chloride, a type of heavy-duty plastic – for interior piping. In some places even pipes coming off municipal water mains can be made of lightweight materials. It is easier than ever to run pipes anywhere a customer wants them.

The vast selection of plumbing equipment and accessories is also a boon to plumbers. Never before in history have so many people had access to so many choices. Chrome, brass, nickel, oil-rubbed bronze or stainless steel? High-flow or low-flow? Single function showerhead or nine-function showerhead? The choices would be unimaginable to consumers from only a generation ago. The variety of options means that fashions change more rapidly than ever before. For most of human history nobody cared about the "design" of their kitchen or bathroom – they were just glad to have them. Today, it is not unusual for homeowners to update their kitchens and bathrooms every decade or so to keep up with fashion and maintain their property value. Plumbing has come a long way and you can be a part of the next revolution.

WHERE YOU WILL WORK

THE DEMAND FOR SKILLED PLUMBERS is universal. If you pursue a career in plumbing you will be able to work anywhere you want. Your only real consideration is what kind of plumbing you want to do.

The person most people think of when they hear the word "plumber" is the person who unclogs drains and replaces hot-water heaters in their homes. True enough, but there is a world of difference between the plumbing required by a single-family house and that required by a high-rise residential building, a commercial building, a factory, and a public-works project. Have you ever considered the physics involved in flushing a toilet 50 stories off the ground? If you want to specialize in common residential and light commercial plumbing you can work just about anywhere. If you want to learn the ins and outs of high-rise plumbing you will need to set your sights on a career in a big city like New York or Chicago with many tall buildings. Industrial properties can be found just about everywhere and so can public works. You will not need to specialize until much later in your career, but you should be thinking about these things as you get started on your career.

You should also consider opportunities for

entrepreneurship and business ownership. Many plumbers are self-employed. In fact, plumbing is an excellent career choice for handy people who want to be their own boss. Complete an apprenticeship, spend five years or so working for somebody else, and then you can strike out on your own in your early 30s. Take a look at the market for plumbers in the area where you wish to live. Is it growing, creating a demand for plumbers for new homes or commercial buildings? Is it going through a generational turnover in which older folks are moving out and younger dwellers are moving in, creating demand because the new homeowners all want to update their kitchens and bathrooms? Choosing the right metropolitan area at the right time could be your ticket to success as an entrepreneur.

DESCRIPTION OF THE WORK

Apprentice Plumbers

All plumbers start out as apprentices. This is the period in which careerists really learn their craft. More importantly, this is also when they learn if they have what it takes to make a living in the plumbing business.

Most states require a four- or five-year apprenticeship under the supervision of a journey worker plumber. Apprentices are typically assigned to a single plumbing business for the duration of their apprenticeships, and that business may assign a specific journey worker to take responsibility for the apprentice's progress. In practice, the apprentice may go to a job site and work alongside whichever journey worker is available to guide and supervise on that particular job. An apprenticeship is like earning a bachelor's degree. It takes just as long and is every bit as challenging.

Apprentices do not just carry pipe around job sites, either. Apprenticeships are targeted training programs with very specific educational requirements, just like any other training program. Apprentices are expected to master basic plumbing principles before moving on to learn about specific applications. No one just shows up and starts installing an entire kitchen without considerable training and experience.

Apprenticeships typically require apprentices to work a standard 40-hour week and complete about 250 hours of classroom training per year. Classes concentrate on topics like the science and engineering behind plumbing, basic business management, cost estimating, labor relations, and regulatory issues. There is much to learn in the classroom. Some apprenticeship programs are conducted in conjunction with community colleges and can lead to an associate degree as well as certification as a journey worker plumber.

An apprenticeship may sound like a long and tedious way to learn a skill. It is important to remember, however, that apprenticeships have been around for hundreds of years for a reason. They are an excellent way to pass along knowledge from the current generation to the next. That is why apprenticeships are so important to skilled trades like plumbing, electricity, and carpentry. The only way to learn these trades, to really understand the job, is to work alongside an established professional for a few years. Most apprentices remember their apprenticeships fondly and are honored when the time comes to return the favor by supervising new apprentices later in their careers.

Journey worker Plumbers

Most plumbers are journey workers. Historically the title was "journeyman," and the old title is still in use in many jurisdictions, but journey worker has been adopted in most places as a gender-neutral descriptor for the most fundamental position within the hierarchy of professional plumbers.

A journey worker, simply stated, is a plumber who has completed an apprenticeship and earned a license to work alone and without supervision. Most plumbers are journey workers. If a team of plumbers arrives at a construction site, the team will probably include a couple of apprentices, at least one master plumber, and a number of journey workers. Many plumbers remain at the journey worker level for their entire careers. Journey workers usually have minimal managerial responsibilities. They are eligible for overtime and get to spend their days doing plumbing rather than shuffling papers. They do not take their work home with them, either.

Most journey workers are generalists who can be assigned to just about any plumbing job. These can range from simple tasks that many homeowners handle themselves, like unclogging a drain, to more sophisticated projects like running pipes for an addition or for new construction. These jobs can include a wide variety of challenges, many of which are unrelentingly physical. Lying on your back underneath a sink is uncomfortable. So is reaching around into tight spots to find connections or valves. Nobody likes spending the day in a dirty, spider-infested crawlspace, or digging a trench across a yard to bury a pipe leading to a water main. Journey workers often end up doing some landscaping because they have to replace anything they remove when they dig trenches. This can be a very dirty job.

It can also be a dangerous one. Plumbers are not at risk

of electrocution the way electricians are, nor are they at risk of losing a finger in a table saw like carpenters are. They are at significant risk of repetitive stress injuries and ordinary pulls and strains. Journey worker plumbers are unlikely to suffer serious bodily harm on the job but the aches and pains never end.

Master Plumbers

Master plumbers are at the top of the plumbing hierarchy. In most jurisdictions plumbers must serve for a minimum period of years, typically seven or more, before they can take the examination to become a master plumber. Some jurisdictions require anybody with a master plumber license to also become a licensed contractor and carry liability insurance.

Master plumbers take charge of journey workers. Master plumbers are assigned to manage large teams on complex projects. This can include inspecting the work of journey workers, general scheduling and management, reviewing blueprints and even handling the design of complex plumbing systems.

Many plumbing contractors are master plumbers. Most jurisdictions do not require master plumber certification in order to become a licensed plumbing contractor, but there is no doubt that being able to tell potential clients that you are at the highest level of the profession is a valuable marketing tool. It also helps to attract the best journey workers to your crews.

It is not unusual for master plumbers to also earn bachelor's degrees in engineering. Designing the plumbing for a major commercial building, for example, is not the kind of thing anybody learns through an apprenticeship. If you really want to rise to the top of the plumbing profession you should set your sights on becoming a master plumber.

Plumbing Contractors

Most plumbers work for plumbing contractors but relatively few are contractors in their own right. Contractors are business owners who have paid fees and passed examinations to earn licenses to market their services to the public. Becoming a licensed contractor is similar to becoming a licensed plumber but the two paths are technically distinct.

Journey worker plumbers, for example, are licensed to work without supervision. That means they have proven that they have the expertise necessary to take on all but the most specialized plumbing jobs – for which they would be required to have a very specific certification – and be likely to succeed. Licensing is important. It means that a journey worker has been tested, possesses a specific set of skills, and can reasonably be expected to complete a job correctly and in a reasonable amount of time.

Journey workers are licensed to work for contractors. Contractors are licensed to work for everybody. Journey workers who wish to open their own plumbing contracting companies must pass an exam and pay a fee to become a licensed contractor in addition to maintaining their licensure as a journey worker.

Many plumbing contractors are master plumbers. Only the most accomplished plumbers become master plumbers, and only the most ambitious plumbers become contractors.

Becoming a plumbing contractor comes with its own advantages and pitfalls. Being your own boss is great. You may make a lot of money. You may be able to take vacation time. You may be called for emergencies at all hours of the day and night. You may build a very lucrative business and then lose it when a recession hits and homeowners stop renovating their kitchens and bathrooms. All plumbing contractors start their careers by working for somebody else. You have plenty of time to think about the possibilities of becoming a contractor.

Related Careers

Plumbing is closely related to pipefitting, the profession that specializes in designing and building piping for systems other than conventional plumbing. Plumbers deal almost exclusively with water, while pipefitters deal with gas, steam and specialized water systems like fire sprinklers. Their commonality is that they all do cutting, bending, and routing of pipe. Many residential plumbers also handle simple jobs involving gas lines, for example, because in most residential applications any routing of pipe is similar. There is no need to call a specialized pipefitter to fix a simple gas valve in a residential kitchen.

There is definitely a need to call specialized pipefitters to design, build and maintain the maze of pipes in an industrial plant or a major commercial building. Search the internet for photos of oil refineries, for example, and you will see a mass of pipes! Some pipefitters start out as plumbers and switch to pipefitting later in their careers. Common industrial specialties for pipefitters include gasfitting and steamfitting, both of which are crucial to industrial productivity.

Sprinkler fitters design, build and maintain the piping for fire sprinklers commonly found in business and public buildings and, more and more often, private homes. Sprinkler fitters often work alongside regular plumbers. Some plumbers also work in sprinkler fitting, especially for residential applications. Some careerists also build and maintain residential lawn sprinkler systems, which is another interesting spin on conventional plumbing. Lawn sprinkler specialists need to earn many of the same certifications as their counterparts who spend most of their time working indoors.

STORIES OF PLUMBING PROFESSIONALS

I Own a Plumbing Contracting Company

"I always wanted to be my own boss. That's the main reason I got into the plumbing business. I now own a plumbing contracting company with more than 20 full-time employees. It took years of effort to get to where I am today but I wouldn't have it any other way.

Like many careerists in the skilled trades I started out taking shop classes in high school. I took all of them: carpentry, metalworking and auto mechanics. Metalworking was one of those classes that taught a wide variety of skills with machine tools, while carpentry and auto mechanics taught me all about hand tools. I wasn't sure what I wanted to do after high school, so when a friend suggested plumbing I decided to check it out.

I really liked the idea of a long apprenticeship under the guidance of a journey worker plumber. That's the best way to learn a craft like plumbing. Book learning is essential, and my apprenticeship required 250 hours per year of classroom study, but the only way to really learn this trade is by doing it.

I worked for several plumbing companies as a journey worker. I worked both as a full-time employee and as an on-call freelance plumber, but mostly it was full time. I learned a lot about running a business while I was an employee. I could step back and observe when I wanted to, and I could also step up and take charge when I wanted to. Working for somebody else for a few years is really the only way to learn the things you need to know to start your own business.

When I decided to start my own company, one of the first things I did was take the exam to become a certified master plumber. Being a master plumber allows me to take charge of any plumbing project of any size. I can prepare blueprints, hire and manage as many plumbers as I need and take responsibility for everything.

I built my business slowly, starting with only myself and a van filled with tools. After about a year I was able to hire an assistant, and then we affiliated with the local apprenticeship program and started taking on apprentices. We started out specializing in residential and light commercial contracting but eventually branched out into larger, more complex projects like industrial plumbing and multi-unit residential projects like large apartment buildings. I hired new journey workers as fast as I could. I quickly learned, however, that employing a large staff requires maintaining a full schedule of plumbing jobs. I can't afford to pay plumbers who aren't working. So instead of full-time employees, I resort to subcontracting some work to smaller plumbing contractors. It's a common practice and one of many things you have to learn in order to run your own business profitably.

I've had to learn a lot of other things, too. I tried to do my own accounting in the beginning. Big mistake. I hire an accounting firm now. It's also good to keep a lawyer on retainer to review major contracts, some of which are far too complex for me to sort out.

Marketing and advertising I seem to do pretty well on my own. My teenage daughter runs my website and Facebook page, which keeps that function in the family, at least. I've taken a few business classes at the local community college, too. I recommend it for anybody who thinks they have what it takes to run their own business. You may have the desire but that doesn't mean you have the skills. Take the time to learn and you will be rewarded."

I Am a Plumber's Apprentice

"I've been a plumber's apprentice for almost two years and it amazes me how much I still have to learn.

I didn't always know I wanted to be a plumber. I just knew I liked to work with my hands, wasn't crazy about sitting behind a desk, and wanted to learn a trade that was fairly recession proof and that I would enjoy. Plumbing entered the picture when I saw a sign for a seminar on an apprenticeship program at the community college where I was taking shop classes. The college administers this program in conjunction with the local union. It takes four years and requires 40 hours of work per week and 250 hours of classroom training per year. You can take the program that results in journey worker status, or the more time-consuming version that also results in an associate degree in business administration. I chose the degree option. Graduating is still a couple of years off but I'm glad I'm putting in the extra time now while I'm still relatively young and don't have many commitments.

A typical day for me starts at about 7 a.m. with coffee with my supervisor and the rest of the team. Then we head out to our jobs. Some jobs take several days or even weeks, while others are one-day jobs. We start working as soon as we arrive. I generally get the grunt work like carrying tools from the van to the work site, but I'm also at the point where I can do much of the work alongside my trainer. I still watch carefully because even a small mistake could be devastating. I've seen small leaks turn into floods because somebody did something stupid. That's why apprenticeships take so long. You can learn the basic skills in a year or two. It takes another couple of years to learn the little things that keep you out of trouble.

To give an example, I once installed a bathtub water faucet incorrectly. Most valves nowadays come pre-assembled. They are attached to a pipe that reaches down a few inches to connect to a faucet. The pipe flexes a little, so it has to be secured in place with a strap. I forgot to install the strap. After we finished the installation another team installed drywall over the valve assembly and then a layer of tile on top of that. When we returned to attach the faucet and valve handles we couldn't stop the faucet from wiggling. We attached it securely to the pipe but the pipe wasn't attached to anything because I had forgotten to install the strap. The only way to fix it was to cut a hole in the tile and drywall and start over. A little mistake turned into a big problem. Thankfully, that's the kind of mistake you only make once.

I'd recommend a plumbing apprenticeship to anybody who really enjoys working with their hands, getting out and about, and taking pride in a job well done. My goal is to start my own plumbing company someday. I'm off to an excellent start."

I Am a Journey worker Plumber

"I've been a journey worker plumber for a decade. I wouldn't do anything else. There is something about the challenge, creativity and freedom of this job that appeals to me.

My state requires a five-year apprenticeship. I enjoyed every minute of it. I am also the first to say that a long apprenticeship is absolutely necessary to succeed in such a complex trade. Think of your apprenticeship as going to college. It takes just as long and is every bit as challenging.

In fact, one of my greatest pleasures as a journeyworker is the opportunity to train up-and-coming apprentices. Some journeyworkers look at apprentices as time-wasting obstacles. I see their point, but that's being cynical. Your job isn't all about work, work, work. Helping to bring up the next generation is important. I've also learned from my apprentices. They look at things with fresh eyes and ask questions I never would have thought to ask. When you become a journey worker, don't avoid apprentice training. It's one of the most satisfying things you'll ever do.

As a journey worker I am one of the small army of plumbers who can work alone, without supervision. That means my boss can send me out on a job and be certain that I will do the job right and make my company look good. I am currently employed full time by a small plumbing company. We take on our own jobs for residential and light commercial clients and subcontract ourselves out for bigger jobs. We just finished a huge job on a new subdivision, for example. We were one of about a dozen plumbing contractors hired to take on the plumbing for a new subdivision of more than 300 houses. As far as we were concerned we were doing the kind of residential work that is our bread and butter. Our little company never could have handled the entire project, however.

I like residential jobs the best. The work is often very simple, like unclogging drains or replacing faucet cartridges, but I like these jobs because they get me out and about in the community. I meet people and help them out. People are often very grateful when I solve their problems, and that makes me feel good. Even a problem with the easiest solution may have been a huge worry for the homeowner. I also hand out a lot of business cards and get plenty of word-of-mouth referrals this way.

I sometimes think about opening my own company but I'm not sure. This is a pretty good job as it is and I'm not sure I need the managerial headaches that would come with having my name on the side of the van."

I Am a Master Plumber for a Large Construction Company

"I am a licensed master plumber with a bachelor's degree in civil engineering. I started out as a plumber's apprentice, just like all plumbers. I completed a five-year program that also resulted in an associate degree in general studies that I was able to transfer to a four-year university to earn a bachelor's degree in civil engineering. I worked part time as a journey worker plumber while going to college, which was great. Many of my college friends were doing jobs like delivering pizzas but I was making real money doing a real job. During my studies I completed an internship with a very large construction company. The kind of company that builds skyscrapers. I was intrigued and decided to go to graduate school.

I kept up the same pace in graduate school, working as a journey worker plumber and juggling my class schedule to fit around my work. I took my master plumber exam shortly before I finished graduate school, resulting in a very impressive résumé.

I went to work for the company where I completed my internship and I'm still there today. As a master plumber I am authorized to modify and contribute to blueprints for entire plumbing systems. As a civil engineer I have additional structural knowledge that helps me to solve the puzzles that plumbing always presents. Pumping water up and down a 50-story building is very, very complicated. To make that happen properly I need multiple credentials and years of experience.

I love my career because I am at the pinnacle of my profession. I can handle all aspects of plumbing for high-visibility projects with budgets in the hundreds of millions of dollars. I am very proud of what I do."

PERSONAL QUALIFICATIONS

BEING A PLUMBER CAN BE HARD work, and there are many things to learn. There are also some qualities you should already have if you really want to be successful in a plumbing career.

First, you should be mechanically inclined. That is a broad term used to describe a knack for being able to build or fix things like cars, houses, appliances, and plumbing systems. Mechanically inclined people who do not pursue a mechanical profession take great pride in calling themselves "do-it-yourselfers" because they are always willing to take a crack at a household fix-it project themselves, often with excellent results. Your innate mechanical skills can lead you to a rewarding career as a professional plumber.

Do you tinker with cars? Fix squeaky doors, sticky windows, and leaky toilets? Do you like to fix appliances that have mysteriously stopped working? Did you build a clubhouse when you were a kid? If you can answer yes to at least a few of these questions you may have the mechanical inclination necessary to succeed as a plumber. The truth is, most people are not mechanically inclined and would not be suited to earning a living as a plumber.

Plumbing is not a desk job. Even if you start your own company and spend most of your time directing the work of others, you will still find yourself doing very physical jobs from time to time. Entry-level plumbers spend most of their time lifting, bending and stretching, often in uncomfortable areas like under sinks or in crawl spaces. They have to lug pipes from trucks to job sites and carry heavy tools to wherever they are needed. Sometimes they have to dig trenches and replace landscaping. They may also have to open up walls and cut through studs and other impediments. This is hard physical labor. If you become a plumber you will get your hands dirty every day.

Basic business skills are essential to a successful career in plumbing. Even as an entry-level plumber with little to no responsibility for negotiating contracts, buying supplies, or otherwise scheduling work you will still be expected to provide excellent customer service. Not everybody has a knack for this. Leaving the customer with a good impression is what makes a big difference in word-of-mouth referrals. Customers who have a good experience with you will tell their friends and neighbors, which will result in more business. Customers who have a bad experience will also tell their friends and neighbors, and that could cause you to lose business. Today, a recommendation (or complaint) on Yelp and Angie's List can make or break your business. If you plan to move up in the world and start your own business you will need to learn more advanced business skills like accounting, management and marketing. Many successful plumbers earn degrees in business administration in order to gain an edge on the competition.

ATTRACTIVE FEATURES

CAREERS IN PLUMBING CAN BE VERY satisfying. People will always need plumbers, which creates excellent job security.

Ask people in skilled trades when they knew what they wanted to do and most will tell you that they always knew they wanted to do something with their hands. It may have taken a few years and some experimentation to decide on plumbing or carpentry or auto mechanics, for example, but very few people in these professions will ever tell you that they really wanted to sit behind a desk in an office, but wound up getting their hands dirty. Most skilled tradespeople would not want to do anything else. If you are one of these lucky people you will never regret your choice of career.

Plumbing is a necessity in modern society. Even the most rustic campers still like to come home to a hot shower and a proper toilet after a few days in the woods. This creates excellent job security for skilled plumbing careerists. There are also some other factors working in your favor. Plumbing is a key element in modern public health programs. Epidemics are caused in much of the world by inadequate municipal plumbing systems that lack modern safety features. Diseases are rarely transmitted via drinking water in modern societies with up-to-date plumbing. Plumbing codes in American cities are very strict and not optional. This also keeps plumbers employed.

At the other end of the spectrum is the fact that nothing adds to the value of a home like new kitchens and bathrooms. Homeowners will always be willing to spend big money on upgrading their kitchens and bathrooms – where the plumbing is – in order to boost their property values. Plumbers are in demand for these jobs.

Most plumbers are either self-employed or work for small businesses in which they are only a notch or two removed from the owner. Opportunities for entrepreneurship are almost endless. It is common for plumbers to spend their first few years serving an apprenticeship, then a few more working for somebody else, then start their own business, often with a partner or two they met while on the job. In fact, most small plumbing businesses get started this way. If you think that you would like to be your own boss someday, plumbing is an excellent way to get there.

UNATTRACTIVE ASPECTS

CHIEF AMONG THE DRAWBACKS IS the perception of plumbing as a dirty job for people who cannot go to college. Plumbing is one of those careers often looked down upon by people who fancy themselves superior to those who spend their days soaking in fetid water under other people's sinks. Many people feel the same way about carpenters, auto mechanics, bricklayers, carpet and tile installers, heating and air-conditioning professionals, and just about anybody else whose profession requires them to do dirty, backbreaking chores that most people cannot or will not do for themselves. This is nothing but snobbery and you will just have to ignore it. If all the world's Shakespeare scholars disappeared tomorrow very few people would notice. If all the world's plumbers – and auto mechanics, carpenters, bricklayers, and heating and air-conditioning professionals – disappeared tomorrow the whole world would take notice quickly and be in crisis mode. Take pride in being a plumber, and making good money for doing a very valuable and important job.

While there is pride to be taken in doing hard jobs there is no denying the fact that hard work takes its toll. You will not find very many active plumbers in their 60s, mostly because they have retired or concentrate on management rather than plumbing. Crawling around under sinks can give anybody aches and pains, and it only gets worse as you get older. Lugging heavy pipe around job sites is not fun. Neither is digging trenches to lay pipe or moving around bushes and other landscaping. Being doused in smelly water filled with who-knows-what can eventually get to you. Always going to the drive-thru for lunch because you know you look messy and smell worse is not exactly an esteem-booster. Plumbing is a hard job that may look like a challenge in your 20s but more like a burden later in life.

Many plumbers stay on the same journey worker rung of the plumbing career ladder pretty much from the time they finish their apprenticeship until the time they retire. If you do not take the plunge and start your own business or become a master plumber you may find yourself stuck in a rut for a very long time. Rules vary from one jurisdiction to the next, but some union hierarchies recognize seniority at the expense of skill. Even if you are an excellent plumber, you may not be able to move up in the local pay scale until the older plumbers ahead of you retire. Job security should still be good, but your paycheck may not change much for years at a time.

EDUCATION AND TRAINING

BECOMING AN EXCELLENT PLUMBER requires many years of education and training. Do not think for a second that plumbing is easy just because plumbers do not have to go to college. Plumbing just requires a different kind of education.

You can start your plumbing education while you are still in high school. Take as many shop classes as you can, and get good grades. Do not neglect regular academic subjects like math and English. You will need excellent communication skills to deal with customers, and excellent math skills to untangle plumbing problems. There is no excuse not to do your best in high school. Being an excellent student will become a habit for which you will be grateful later in life.

All plumbers start their careers as apprentices working under the supervision of experienced professionals, which is an excellent way to learn a trade. Regulations vary from one state to the next. Most apprenticeship programs require four or five years of work experience under the supervision of journey worker plumbers plus extensive classroom training. A typical apprenticeship program requires 2000 hours of work per year plus another 250 hours of classroom study per year. Completing an apprenticeship is very much like earning a bachelor's degree except that "homework" takes 40 hours per week and comes with a paycheck. A small paycheck, to be sure, but apprentices are paid a salary.

After completing an apprenticeship you will become a journey worker plumber. Legally, this means you can work independently, without supervision from a more-senior plumber. Most plumbers stay at the journey worker level for the rest of their careers, moving up the pay scale in terms of seniority and maybe picking up a few certifications along the way. Some become master plumbers, which requires working as a journey worker for many years – the exact number depends upon the state in question – and passing an exhaustive examination. You may decide to become a master plumber someday, but that is not a decision you have to worry about right now.

Many unions and professional associations offer professional certifications for journey worker plumbers. Certifications can be earned by passing an exam, putting in a required amount of time in a specialty, or both. Professional certifications can help journey worker plumbers to tailor their résumés to the kind of work that interests them the most. This can boost their paychecks and their enthusiasm, both of which are important.

How much education you pursue beyond an apprenticeship and professional certifications is up to you. If you have your sights set on opening your own business you should seriously consider earning a degree in business administration. Most apprenticeship programs will touch upon basic business skills during your training but if your goal is to be your own boss you owe it to yourself to get a solid education in business. Many community colleges offer associate degrees in business administration that are tailored for careerists who want to run small businesses. You can also go all the way and earn a bachelor's degree in business administration.

EARNINGS

THE ONLY WAY TO GET RICH AS A plumber is to become a plumbing contractor handling major commercial or industrial projects. On the other hand, few working plumbers fail to maintain a good living. A solid paycheck is one of the advantages of pursuing a career in something as essential as plumbing.

Differences in union rules and among different regions of the United States create some disparities in income, but it is safe to say that most full-time journey worker plumbers well past their apprenticeship years earn about $55,000 to $65,000 per year, a figure comparable to many trades. Very senior master plumbers with multiple certifications and who work in high-cost areas can earn as much as $100,000 per year. The sky is the limit for self-employed plumbers, and especially for those who build sizable businesses and take on large commercial jobs. Some determined entrepreneurs become millionaires in the plumbing business.

You will not be paid so well when you start out, however. Apprentices can be paid as little as 30 percent of the prevailing wage for fully-qualified plumbers. Keep in mind the fact that apprentices are being trained by professionals who have to take time away from doing their work to assist and supervise the apprentices. This costs money and takes time. Also, it takes several years of hands-on experience to gain the knowledge necessary to become an excellent plumber, during which time apprentices cannot command the kind of wages earned by their senior counterparts.

Most plumbers belong to a union. The largest plumber's union is the United Association of Plumbers, Fitters, Welders and Service Techs, commonly known as the UA, which has local affiliates across the country. There are also local and regional unions. Unions tend to set levels for wages and other forms of compensation like health insurance, paid time off, and disability insurance, which is a critical consideration for tradespeople who are at risk of being injured on the job. Unions also set rules for overtime pay, which can be very lucrative. In most jurisdictions, plumbers are paid time-and-a-half for all hours worked past 40 hours per week. It should come as no surprise that many plumbers routinely work 50 hours per week, which boosts their annual income considerably. Plumbers may also be paid a higher rate for responding to emergencies, especially late at night.

OPPORTUNITIES

NOW IS A GOOD TIME TO GET INTO the plumbing business. Over the next decade or so demand for plumbers is expected to grow more rapidly than the demand for most other careers, which is good news for you. You can also move up in the world by earning additional certifications or starting your own business.

Plumbers of all types are in great demand for several reasons. New home construction creates demand for plumbing professionals to install kitchens, bathrooms and the bits of plumbing that most people never think about, like drains and water heaters. Existing homes will always require maintenance. Since nothing boosts the value of a home more than upgraded kitchens and bathrooms there will always be demand from homeowners eager to increase their property value.

The UA offers training programs for plumbing careerists at every stage of their career. After completing an apprenticeship and becoming a journey worker, plumbers can build their résumés by earning certifications in specialties like backflow technology and plastic piping installation, or even earn UA Star certification by learning how to install complex piping for important systems like building fire sprinklers. Plumbers do not have to limit themselves to one type of plumbing.

The biggest opportunity for advancement in the plumbing business is your own ambition. Like most trades, plumbing is ripe for entrepreneurship. Starting your own business is not easy but it is not an impossible dream, either. Many entrepreneurial plumbers spend a few years working for somebody else in order to learn as much as possible before they strike out on their own. Some take classes and earn associate or even bachelor's degrees in business administration in order to gain the business skills they may not learn while working for somebody else. Many entrepreneurs are happy to build small businesses that specialize in residential and light commercial plumbing projects. They employ a handful of plumbers and an administrative assistant and, if they are smart, pay outside accountants and lawyers to keep them out of trouble. Really ambitious plumbers may build large businesses specializing in municipal and large commercial projects. This kind of business is very capital intensive and can require millions of dollars in heavy equipment.

GETTING STARTED

SO YOU HAVE SET YOUR SIGHTS ON A career in plumbing! The first thing you need to do is apply to an apprenticeship program. After becoming a journey worker you can set your sights on becoming a master plumber or a licensed contractor. For now, however, you need to keep your eyes on the prize and get into the business.

Getting into an apprenticeship is a fairly simple process. Most unions require prospective apprentices to fill out an application and sit for an interview, just like an applicant for any other job. Decision-makers will look for candidates with good grades in high school and, preferably, at least a little experience that indicates they will do well in the plumbing trade. Good grades in shop classes will fulfill this requirement, as will part-time jobs in plumbing or a related trade. Selection is not automatic. You will have to compete for available positions. Once selected, the hard work begins. Take your apprentice years seriously! You will learn a great deal, and, just as importantly, you will start building a professional reputation that could follow you for the rest of your career. Very few professions offer multi-year training programs under the supervision of senior careerists, and with a regular paycheck in addition to the free training. Take advantage of this unique time in your life.

Most plumbers spend their careers as journey workers, which means they can work on their own without supervision. A small number of plumbers become master plumbers. In most states becoming a master plumber requires applicants to spend a certain number of years as a journey worker and then pass an examination. Some states also require master plumbers to obtain licenses to work as independent contractors, whether they intend to do so or not. There are also various insurance requirements for master plumbers who will be responsible for supervising large numbers of plumbers and managing complex projects. There is no rule saying that you have to set your sights on becoming a master plumber. Being a reliable, hard-working journey worker comes with good earnings and plenty of job satisfaction. But if you want to move up a rung you have options.

For now, however, your primary goal is to get into the business. Get good grades in high school, take shop classes, get a little experience in plumbing and other trades that require excellent manual skills. Be ready to put your best foot forward when the time comes to apply for an apprenticeship. Once you have the apprenticeship in place, stay focused. Four to five years is a long time to spend in training, and some apprentices lose their patience and drop out of their program. Do not let this happen to you. Plumbing is an excellent career choice with plenty of options for future growth.

ASSOCIATIONS, PERIODICALS, WEBSITES

■ **American Backflow Prevention Association**
www.abpa.org

■ **American Society of Plumbing Engineers**
www.aspe.org

■ **American Society of Sanitary Engineering**
www.asse-plumbing.org

■ **American Water Works Association**
www.awwa.org

■ **California Apprenticeship Coordinators Association**
www.calapprenticeship.org

■ **Chicago Pipefitters Local 597**
www.pf597.org

■ **Contractor Magazine**
www.contractor.com

■ **DIY Plumbing Advice**
www.diyplumbingadvice.com

■ **Everest College**
www.everest.edu

■ **Faucet Central**
www.faucetcentral.com

■ **Grainger Industrial Supply**
www.grainger.com

■ **Illinois Plumbing Inspectors Association**
www.ipiassoc.org

■ **International Association of Plumbing and Mechanical Officials**
www.iapmo.org

■ **International Code Council**
www.iccsafe.org

■ **National Kitchen and Bath Association**
www.nkba.org

■ **Penn Foster Career School**
www.pennfoster.edu

■ **Plumbing and Mechanical Magazine**
www.pmmag.com

■ **Plumbing, Heating, Cooling Contractors of California**
www.caphcc.org

■ **Plumbing Manufacturers International**
www.safeplumbing.org

■ **Plumbing Mart**
www.plumbingmart.com

■ **Plumbing Net**
www.plumbingnet.com

■ **Plumbing Web**
www.plumbingweb.com

■ **PM Engineer Magazine**
www.pmengineer.com

■ **Stratford Career Institute**
www.scitraining.com

■ **United Association Union of Plumbers, Fitters, Welders and Service Techs**
www.ua.org

■ **Water Quality Association**
www.wqa.org